JEWELRY APPRAISAL & GEMS

MUST KNOW BEFORE APPRAISAL

EDUARDO ALCANTARA CG CDS

Eduardo Alcantara CG CDS

JEWELRY APPRAISAL & GEMS

MUST KNOW BEFORE APPRAISAL

Copyrights © First edition 2023

Autor: Eduardo Alcantara CG CDS

alcantarae9@gmail.com

TO:

PUBLIC IN GENERAL

This book it is to inform the public as introduction to answer the more commons questions, they have related about what to expect when they want or need a GEM JEWELRY APPRAISAL.

Special dedication to **International Gem Society (IGS)** and membership around the world for the support to professionals and enthusiasts in gemology sciences.

To all the members of my family.

CONTENTS

CHAPTER 1

Introduction to Jewelry Appraisal: An overview of what jewelry appraisal is, why it is important, and the types of appraisals.

Jewelry appraisal is the process of finding the value of a piece of jewelry. It is an expert evaluation that typically includes a detailed description of the jewelry, including its metal content, gemstones, and any other relevant characteristics. The appraisal also includes an estimated value of the jewelry, which may be based on factors such as current market conditions, the rarity and quality of the materials used, and the overall quality of the piece.

Why is Jewelry Appraisal Important?

Jewelry appraisal is important for a variety of reasons. One of the most common reasons is for insurance purposes. Having a jewelry appraisal can help ensure that your jewelry is properly insured in case of loss or damage. Jewelry appraisals are also important for estate planning. They can help find the value of assets for tax, distribution purposes, and as financial collateral. Additionally, jewelry appraisals can be helpful when selling jewelry. They can provide potential buyers with an independent assessment of a piece's value.

Different Types of Appraisals:

There are distinct types of jewelry appraisal, depending on the purpose and scope of the appraisal. Here are the most common types of jewelry appraisal:

Insurance appraisal: An insurance appraisal is typically performed to decide the value of a piece of jewelry for insurance purposes. This type of appraisal takes into account the current market value of the jewelry, as well as any historical or sentimental value.

Estate appraisal: An estate appraisal is performed to determine the value of a piece of jewelry for estate planning or probate purposes. This type of appraisal may take into account the current market value, as well as any historical or sentimental value, and may involve research into the piece's provenance and history.

Resale appraisal: A resale appraisal is made to decide the current market value of a piece of jewelry for resale purposes. This type of appraisal takes into account the condition of the jewelry, as well as current market trends and demand.

Donation appraisal: A donation appraisal is performed to determine the fair market value of a piece of jewelry that is being donated to a charitable organization. This type of appraisal is often required by the IRS for tax purposes.

Liquidation appraisal: A liquidation appraisal is performed to determine the fair market value of a piece of jewelry that is being sold in a liquidation sale or auction. This type of appraisal considers the condition of the jewelry, as well as current market trends and demand.

Probate appraisal: is an appraisal before to travel out the country with high valued jewelry to probate to customs when arriving back of its existence inside the country before travelling.

In general, a jewelry appraisal will involve a detailed examination of the piece of jewelry, including an assessment of its materials, artisanship, signature and condition. The appraiser will then determine the appropriate value for the jewelry based on the type of appraisal being performed. It is important to choose a qualified and experienced appraiser to ensure an accurate and reliable appraisal.

In conclusion, jewelry appraisal is an important process for finding the value of a piece of jewelry. The Appraisal Report it is used for insurance, estate planning, divorce cases, inheritances, finance collaterals, curiosity or selling jewelry. Understanding the different types of appraisals, the factors that influence value, and the process of getting an appraisal can help you make informed decisions about your jewelry.

CHAPTER 2

The Appraisal Process: A step-by-step guide to the process of having jewelry appraised, including what to expect and what information is needed.

1. **Finding an Appraiser**: The first step in the appraisal process is finding a qualified professional Gemologist appraiser. Look for an appraiser who is a member of a professional organization. In USA, developed countries and some others you can find it one or more of this kind of institutions. You can also ask for recommendations from jewelers or other professionals in the industry.

2. **Scheduling an Appointment**: Once you have found an appraiser, schedule an appointment to have your jewelry appraised. Be prepared to provide information about the jewelry, such as how and when you acquired it and any documentation you have as invoice, ID Report, provenance, or previous appraisal.

3. **The Appraisal Examination**: During the appointment, the appraiser will examine your jewelry in detail. This will include a visual inspection, as well as testing for the metal content and any gemstones. The appraiser may also take photographs of the jewelry for the appraisal report and can evaluate in your presence.

4. **Report Preparation**: After the examination, the appraiser will prepare a report that includes a detailed description of the jewelry, including its metal content, gemstones, and any other relevant characteristics. The report will also include an estimated value of the jewelry.

5. **Review and Signature**: Once the report is complete, the appraiser will provide you with a copy for review and signature. Make sure to read the report carefully and ask any questions you may have.

6. **Finalizing the Appraisal**: Once you have reviewed and signed the report, the appraisal process is complete. You can use the report for insurance or estate planning purposes or as a reference when selling the jewelry.

It is important to note that the process and time limit of an appraisal may vary depending on the appraiser, the type of appraisal and the complexity of the jewelry. One common complexity evaluating stones in jewelry is the relative as the mounting permit. It is recommended to confirm with the appraiser before the appointment as normally these services are made. Ask any questions you may have before the appraisal.

Each appraiser has own fees depending of the complexity of the jewelry, if the jewelry has previous ID report, the experience, specialties, if need to travel to customer place, geographic, market, reputation and others factors. Also, in occasions is necessary a fee advance.

In conclusion, the jewelry appraisal process typically involves finding a qualified appraiser, scheduling an appointment, signing contract, examination of the jewelry, report preparation, review, and signature, and finalizing appraisal. Knowing what to expect.

CHAPTER 3

Understanding Gemstones: A guide to identify and evaluating several types of gemstones, including diamonds, sapphires, emeralds, and more.

Gemstones are precious or semi-precious minerals used in jewelry making. They are valued for their beauty, durability, and rarity. The most known gemstones include diamonds, sapphires, emeralds, rubies, and pearls. In this guide, we will take a closer look at some of the most popular gemstones and what you should know about identifying and evaluating them.

Diamonds: Diamonds are the most popular and valuable gemstone. They are one of the most valuable gemstones in the world. They are a type of mineral called carbon, formed deep within the earth's mantle under high temperature and pressure. Diamonds are known for their exceptional hardness and brilliance, which make them prized for use in jewelry and other decorative objects.

Diamonds are graded based on the "four Cs": carat weight, color, clarity, and cut. Carat weight refers to the size of the diamond, with larger diamonds being more valuable. Color grades range from D (colorless) to Z (light yellow or brown), with colorless diamonds being the most valuable. Clarity grades range from Flawless (no internal or external blemishes) to Included (visible blemishes), with higher clarity grades being more valuable. Cut refers to the diamond's shape and how well is cut and polished to enhance its brilliance.

Diamonds are used in a variety of jewelry, including engagement rings, earrings, necklaces, and bracelets. They are also used in industrial applications, such as cutting tools and drill bits, due to their exceptional hardness.

Diamonds are primarily mined in Africa, Russia, Canada, and Australia. The diamond trade is subject to regulation by the Kimberley Process Certification Scheme, which seeks to prevent the trade of "conflict diamonds" (diamonds that have been mined and sold to finance armed conflict).

Diamonds are often considered a symbol of love and commitment, and they are a popular choice for engagement rings and other special occasions. They can also be a valuable investment, with prices for high-quality diamonds often increasing over time.

Sapphires: come in a wide range of colors, with blue being the most popular. Other colors include pink, yellow, and green. Sapphire is a precious gemstone that is highly valued for its beauty, durability, and rarity. It is a type of mineral called corundum, which is a crystalline form of aluminum oxide. Sapphires come in a range of colors, with blue being the most popular and well-known. Other colors include pink, yellow, green, purple, and colorless (known as "white" sapphire).

Sapphire is known for its hardness, which makes it highly resistant to scratching and abrasion. It is also highly lustrous and has excellent clarity, which allows it to reflect light and sparkle beautifully. These qualities make sapphire a popular choice for use in jewelry, especially in engagement rings.

Sapphires are typically mined in countries such as Sri Lanka, Madagascar, and Australia. They can be found in Montana and other parts of the United States. The sapphire value is determined by its color, carat, cut, clarity, origin and treatment. The most valuable sapphires are those that are vivid in color, free of visible inclusions, and have been cut and polished to maximize their brilliance.

Sapphire has been prized for centuries for its beauty and durability, and it is steeped in history and mythology. In ancient times, sapphire was believed to have healing powers and was associated with wisdom and purity. In medieval Europe, sapphire was good to protect against harm and envy.

Overall, sapphire is a highly prized and valuable gemstone that is admired for its beauty and durability. Its popularity in jewelry has made it a symbol of love and commitment, and it is a timeless choice for special occasions and celebrations.

Emeralds: Emeralds are a type of beryl and are known for their green color. The most valuable emeralds are those that are a rich, deep green. Inclusions, or internal flaws, are common in emeralds. Emerald is a precious gemstone that is highly valued for its striking green color and rarity. It is a type of mineral called beryl, which is a silicate of beryllium and aluminum. The green color of emerald is due to the presence of chromium and vanadium in the crystal structure.

Emeralds are found primarily in Colombia, Zambia, Brazil, and Zimbabwe, with smaller deposits in other parts of the world. The most valuable emeralds are those that are vivid green in color and free of inclusions (flaws or imperfections visible to the naked eye). Emeralds that are slightly yellowish or bluish in hue are typically less valuable.

Like other gemstones, emeralds are graded based on their color, clarity, cut, and carat weight. The most valuable emeralds are those that are intense green in color, have few inclusions, and have been cut and polished to maximize their brilliance.

Emerald has been prized for its beauty for centuries and has been used in jewelry since ancient times. It was particularly popular in the Art Deco period of the 1920s and 1930s, and it continues to be a popular choice for fine jewelry today.

In addition to its beauty, emerald is also believed to have healing properties and spiritual significance. It has been associated with love, rebirth, and the heart chakra, and is believed to bring harmony and balance to the wearer.

Overall, emerald is a highly prized and valuable gemstone that is admired for its beauty, rarity, and spiritual significance. It is a timeless choice for special occasions and celebrations and is considered a symbol of love and new beginnings.

Ruby: is a precious gemstone that is highly valued for its rich red color and rarity. It is a type of mineral called corundum, which is a crystalline form of aluminum oxide. The red color of ruby is due to the presence of chromium in the crystal structure.

Rubies are found primarily in Myanmar (formerly known as Burma), as well as in other parts of Southeast Asia, Africa, and South America. The most valuable rubies are those that are vivid red in color and free of inclusions (flaws or imperfections visible to the naked eye). Rubies that have a slight purple or brown hue are typically less valuable.

Like other gemstones, rubies are graded based on their color, clarity, cut, and carat weight. The most valuable rubies are those that are intense red in color, have few inclusions, and have been cut and polished to maximize their brilliance.

Ruby is prized for its beauty for centuries and has been used in jewelry since ancient times. It was particularly popular in the Art Deco period of the 1920s and 1930s, and it continues to be a popular choice for fine jewelry today.

In addition to its beauty, ruby is also believed to have healing properties and spiritual significance. It has been associated with vitality, courage, and passion, and to stimulate the heart chakra and promote positive energy.

Overall, ruby is a highly prized and valuable gemstone that is admired for its beauty, rarity, and spiritual significance. It is a timeless choice for special occasions and celebrations and is often considered a symbol of love and passion.

Pearls: Pearls are formed inside certain types of mollusks. They are valued for their luster and iridescence. Natural pearls are very rare and can be quite valuable. Cultured pearls, which are grown in pearl farms, are much more common and are used in most pearl jewelry.

Pearls are unique gemstones that take form inside mollusks such as oysters and mussels. They are made of calcium carbonate in the form of nacre, which is secreted by the mollusk to protect its soft tissues from irritants such as sand or parasites. Over time, layer upon layer of nacre is deposited, forming a pearl.

Pearls come in a range of colors, including white, cream, pink, silver, and black. They can be round, oval, or irregularly shaped, and can vary in size from a few millimeters to over a centimeter in diameter.

Natural pearls are rare and highly prized, as they form completely by chance in the wild. Cultured pearls, on the other hand, are created by human intervention, in which a technician inserts a nucleus into the mollusk to stimulate the growth of a pearl.

Pearls have been valued for their beauty and rarity for centuries, and have been used in jewelry and adornments since ancient times. They are often associated with purity, innocence, and feminine beauty, and are a popular choice for wedding and bridal jewelry.

The value of a pearl is determined by its size, shape, color, luster, surface quality, and rarity. In general, pearls that are larger, rounder, and more lustrous are more valuable. South Sea pearls, which are grown in the warm waters of the South Pacific, are among the most highly prized and valuable pearls in the world.

Overall, pearls are a timeless and elegant choice for jewelry, and are admired for their natural beauty, rarity, and symbolic significance.

Other gemstones: Tourmaline, Topaz, Opal, Tanzanite, Amethyst, Garnet, etc.

When evaluating gemstones, it is important to consider factors such as color, clarity, cut, and carat weight; the origin treatment, signature and historical background are others key factors. The rarer and more desirable the gemstone, the more valuable it will be. It is also important to note that gemstones are often treated to enhance their color or clarity, and this should be disclosed by the seller or appraiser.

In conclusion, understanding gemstones is an important aspect of jewelry appraisal. Each gemstone has unique characteristics that can affect its value. Evaluating the 4 C's, color, clarity, cut, and carat weight are the key factors when evaluating a gemstone. Additionally, knowing about other gemstones, besides the most popular ones, can help you make informed decisions about your jewelry.

To evaluate the 4 C of Diamonds and precious stones as Ruby, Sapphire, Emerald, Jadeite, and many others the gemologist must use a serial of instruments and technics, usually a Gemological Laboratory in several cases a Mobil one.

CHAPTER 4

Understanding Metals: A guide to ID and evaluating different types of precious metals, including gold, silver, and platinum

Precious metals are used in jewelry making and have been valued for their beauty, rarity, and durability. The metals in jewelry are gold, silver, rhodium, platinum, and palladium. In this guide, we will take a closer look at these metals and what you should know about identifying and evaluating them.

Gold: Gold is a soft, yellow metal that is valued for its beauty and rarity. It is measured in karats, with 24 karat gold is 100 % pure form. The purity of gold is also indicated by its hallmark, which is a small stamp on the jewelry. Gold is a precious metal that has been valued for its beauty, rarity, and durability for thousands of years. It is a soft, yellow metal that does not corrode or tarnish, making it ideal for use in jewelry and coins.

Gold is also a popular investment asset, with investors buying gold as a hedge against inflation or economic uncertainty. In addition, central banks around the world hold significant amounts of gold as part of their reserves.

Gold is mined from the earth in various locations around the world, with the largest gold producing countries being China, Australia, Russia, and the United States. Gold is also recycled from old jewelry, electronics, and other sources.

The price of gold can be quite volatile, as it is influenced by a variety of factors including supply and demand, inflation, and geopolitical events. The price of gold is often quoted in U.S. dollars per troy ounce, with the spot price reflecting the current market price for immediate delivery of gold.

Silver: Silver is a white metal that is also valued for its beauty and rarity. Silver is often alloyed with other metals to make it stronger and more durable. Silver is measured in fineness, with .925 indicating that the silver is 92.5% pure. Silver is a precious metal that has been valued for thousands of years for its beauty, malleability, and electrical conductivity. It is a soft, white metal that does not corrode or tarnish easily, making it ideal for use in jewelry, coins, and other decorative objects.

Silver is also used in a variety of industrial applications due to its excellent thermal and electrical conductivity. Also is utilized in electronics, solar panels, medical devices, mirrors, and other products. In addition, silver has antibacterial properties and is used in some medical applications, such as wound dressings.

Silver is mined in various locations around the world, with the largest producers being Mexico, Peru, and China. It is also recycled from a variety of sources, including electronic waste and jewelry.

The price of silver can be volatile, as it is influenced by a variety of factors including supply and demand, inflation, and economic conditions. The price of silver is in U.S. dollars per troy ounce, with the spot price reflecting the current market price for immediate delivery of silver.

Rhodium: Rhodium is a precious metal of platinum group, usually used for plating jewelry because does no corrode, scratch, dent and retains the luster. Rhodium is a rare and lustrous silvery-white precious metal. It is one of the most expensive metals in the world, and its price can be even higher than that of gold or platinum. Rhodium is extremely hard, highly reflective, and has a high melting point, making it valuable in a variety of industrial applications.

Rhodium is primarily used in catalytic converters for automobiles, where it helps to reduce emissions of harmful pollutants. It is also used in the production of electronic components, such as hard disk drives, and in the glass industry for making optical fibers and mirrors.

Rhodium is typically mined as a byproduct of platinum and nickel mining, with the largest producers being South Africa, Russia, and Zimbabwe. Due to its rarity and the difficulty in extracting it from other metals, the price of rhodium can be extremely volatile, with prices rising sharply in recent years.

Rhodium is not commonly used in jewelry making, but it has been used in small amounts as a plating for white gold and silver jewelry to give them a brighter and more durable finish.

Overall, rhodium is an important industrial metal that plays a crucial role in reducing harmful emissions from automobiles and other sources. Its rarity and high value make it an attractive investment for some investors.

Platinum: Platinum is a white metal that is valued for its durability and rarity. It is denser and heavier than gold and silver. Platinum is measured in fineness, with .950 indicating that the platinum is 95% pure. Platinum is a rare and lustrous silver-white precious metal that belongs to the platinum group of metals. It is denser and more durable than gold and silver, and it is highly resistant to corrosion and tarnishing, making it ideal for use in jewelry and other decorative objects.

Platinum is also used in a variety of industrial applications due to its excellent catalytic and electrical properties. It is use in catalytic converters for automobiles, as well as in the production of chemicals, electronic components, and medical devices.

Platinum is primarily mined in South Africa, Russia, and Zimbabwe, with smaller deposits found in North and South America. Due to its rarity and the difficulty in mining and refining it, platinum is often more expensive than gold or silver.

In addition to its industrial uses, platinum is also popular in jewelry making. It is often used in engagement and wedding rings due to its durability and rarity, and it is also a popular choice for men's jewelry.

The price of platinum it is volatile, is influenced by some factors as supply and demand, financial situations, and geopolitical. The price of platinum is quoted in U.S. dollars per troy ounce, with the spot price reflecting the current market price for immediate delivery of platinum.

Overall, platinum is an important industrial metal that is valued for its unique properties and wide range of applications. Its rarity and beauty also make it a popular choice for jewelry and investment.

Palladium: is a rare and lustrous silvery-white precious metal that belongs to the platinum group of metals. It was discovered in 1803 by English chemist William Hyde Wollaston, who named it after the asteroid Pallas, which had been discovered a few years earlier.

Palladium has a number of unique properties that make it valuable in various industries. It is resistant to corrosion, oxidation, and tarnishing, and it has a high melting point and low density. These properties make it useful in the production of catalytic converters for automobiles, as well as in the electronics industry for making capacitors and other components.

Palladium is used in jewelry making, although it is less popular than other precious metals like gold and silver. It is, however, more affordable than platinum, which is also a member of the platinum group metals.

Palladium prices can be quite volatile, as it is a relatively rare metal that is mined in only a few countries, including Russia, South Africa, and Canada. The price of palladium reached an all-time high in 2020, due in part to increased demand for catalytic converters and a decrease in supply due to the COVID-19 pandemic.

When evaluating precious metals, it is important to consider factors such as the metal's purity, weight, and condition. The higher the purity, the more valuable the metal will be. The weight of the metal also affects its value. It's also important to note that some metals may be plated with a thin layer of precious metal, and this should be disclosed by the seller or appraiser.   In conclusion, understanding metals is an important aspect of jewelry appraisal. Each metal has unique characteristics that can affect its value. Evaluating the purity, weight, and condition of the metal are key factors when evaluating a piece of jewelry. Additionally, knowing about other metals, besides the most popular ones, can help you make informed decisions about your jewelry.

CHAPTER 5

Appraisal Methods: An overview of the different methods used to appraise jewelry, including market comparison, cost approach, and replacement value.

When appraising jewelry, there are used various methods to find its value. These methods include market comparison, cost approach, and replacement value. Each method has its own set of advantages and disadvantages, and the best method to use will depend on the specific piece of jewelry being appraised.

Market Comparison: The market comparison method is also known as the comparable sales approach. This method involves comparing the piece of jewelry being appraised to similar pieces that have recently sold in the market. This method is best suited for pieces that have a readily available market of comparable pieces, such as vintage or antique jewelry.

Cost Approach: The cost approach involves determining the value of the piece of jewelry by estimating the cost to reproduce it. This method is best suited for pieces that are unique or custom-made, as there may not be a readily available market of comparable pieces.

Replacement Value: The replacement value method involves determining the value of the piece of jewelry by determining the cost to replace it with a similar piece. This method is best suited for pieces that have a readily available market of comparable pieces, such as modern or mass-produced jewelry.

When appraising jewelry, it is important to note that the value of a piece is influenced by a variety of factors, such as its age, condition, rarity, and desirability. An appraiser will take all these factors into consideration when finding the value of a piece of jewelry.

In conclusion, there are different methods to appraise jewelry and the best method to use will depend on the specific piece of jewelry being appraised. Market comparison, fair market value, cost approach and replacement value are the most common methods. An appraiser will consider several factors such as age, condition, rarity, and desirability to determine the value of a piece.

CHAPTER 6

Factors Affecting Value: A discussion of the several factors that can affect the value of jewelry, including rarity, quality, signature, and condition.

When appraising jewelry, there are factors that can affect the value of a piece. Understanding these factors is important when finding the value of jewelry. Some of the crucial factors include rarity, quality, and condition.

Rarity: Rarity is one of the most important factors that can affect the value of jewelry. Pieces that are one-of-a-kind or limited in number are considered more valuable than those that are mass-produced. This is particularly true for vintage and antique jewelry, where a piece's age and uniqueness can greatly increase its value.

Quality: Quality is another crucial factor that can affect the value of jewelry. Pieces that are well-made and crafted with high-quality materials are considered more valuable than those that are poorly made or made with low-quality materials. This includes factors such as the quality of the metal and gemstones used, as well as the craftsmanship of the piece.

Condition: Condition also can affect the value of jewelry. Pieces that are in excellent condition are considered more valuable than those that are damaged or in poor condition. This includes factors such as scratches, dents, and missing pieces. Pieces that have been well-maintained over time will be in better condition and thus more valuable.

Other factors that can affect the value of jewelry include age, desirability, and historical significance.

A piece that is older or has historical significance can be more valuable than a newer piece, even if the newer piece is of higher quality. Additionally, a piece that is highly desirable or in fashion can be more valuable than a similar piece that is not as desirable.

In conclusion, there are several factors that can affect the value of jewelry, including rarity, quality, and condition. Understanding these factors is important when determining the value of a piece. Factors such as age, desirability, and historical significance can also affect the value of a piece. An appraiser will consider all these factors when determining the value of jewelry.

CHAPTER 7

Appraisals for Insurance: Information on how to use jewelry appraisals for insurance purposes, including how to choose a coverage amount and what to do in case of a loss.

When it comes to ensuring jewelry, an exact appraisal is essential. A jewelry appraisal will determine the value of a piece and help ensure that it is properly insured. Here are key points to keep in mind when using jewelry appraisals for insurance purposes:

Choosing a Coverage Amount: It is important to choose a coverage amount that accurately reflects the value of the piece. An appraiser will determine the value of a piece and provide a written report that can be used to determine the appropriate coverage amount. Be sure to consider both the replacement value and sentimental value of the piece.

In case of a loss: If a piece of jewelry is lost, stolen or damaged, it is important to have a copy of the appraisal. The appraisal will help establish the value of the piece, which is necessary for making a claim with the insurance company.

Choosing an appraiser: Is important to choose a qualified and experienced Gemologist appraiser. Look for an appraiser who is a member of a professional organization; in the regulatory law for appraisal in many countries around the world are regulations to respect and fellow. It is also important to choose an appraiser who specializes in the type of jewelry being appraised.

Keep the appraisal updated: Jewelry value can change over time due to market fluctuations, inflation, and other factors. It's recommended to have your jewelry appraised every 3-5 years to ensure that the coverage amount is up to date.

In conclusion, jewelry appraisals are essential when it comes to insuring jewelry. An accurate appraisal will determine the value of a piece and help ensure that it is properly insured. It's important to choose a qualified and experienced appraiser, keep the appraisal updated, and have a copy of the appraisal in case of loss, theft, or damage.

CHAPTER 8

Appraisals for Estate Planning: Information on how to use jewelry appraisals for estate planning purposes, including how to divide assets among beneficiaries.

When it comes to estate planning, jewelry appraisals can be a useful tool for determining the value of assets and distributing them among beneficiaries. Here are some key points to keep in mind when using jewelry appraisals for estate planning purposes:

Dividing Assets: Jewelry appraisals is used to determine the value of assets, which can help in the division of assets among beneficiaries. An appraiser will determine the value of a piece of jewelry and provide a written report that can be used to distribute assets fairly among beneficiaries.

Fair Market Value: Estate planning often involves determining the fair market value of assets. The appraiser will determine the fair market value of a piece of jewelry, which is the price that a willing buyer pay to a willing seller without compelling or pressures on both parts. This value can also use for tax and legal purposes.

Choosing an appraiser: It is important to choose a qualified and experienced appraiser. Look for an appraiser who is a member of a professional organization, such as the International Society of Appraisers, the American Society of Appraisers, or the National Association of Jewelry Appraisers. It's also important to choose an appraiser who specializes in the type of jewelry being appraised.

Inheritance Tax: Jewelry appraisals can also be used to determine the value of assets for inheritance tax purposes. An accurate appraisal will ensure that the value of assets is correctly reported for tax purposes.

In conclusion, jewelry appraisals can be a useful tool for estate planning. They can be used to determine the value of assets and distribute them among beneficiaries fairly. It's important to choose a qualified and experienced appraiser and use fair market value for tax and legal purposes. Jewelry appraisals can also be used to determine the value of assets for inheritance tax purposes, divorces and for financial use as collateral in financial institutions.

CHAPTER 9

Selling Jewelry: Tips for selling jewelry, including how to price items and where to sell them.

When it comes to selling jewelry, it is important to understand how to price items and where to sell them to get the best return on your investment. Here are tips for selling jewelry:

Pricing: One of the most crucial factors in selling jewelry is pricing. It is important to have a clear understanding of the value of the piece, which is decided by a jewelry appraisal. Be sure to price items competitively and consider both the replacement value and sentimental value of the piece.

Research: Research the market to see what similar pieces are selling for. This will give you an idea of what price range to sell your jewelry in and help you price your items competitively.

Marketing: Make sure to take high-quality photos of your jewelry and write a detailed description of the piece, including any valuable information such as the type of gemstone, metal, and weight. This will help potential buyers understand the value of the piece.

Where to sell: There are diverse options for selling jewelry. One option is to sell it to a jeweler or pawnbroker, but these options may not be as profitable as selling it to a collector or at an auction.. Another possibility is to sell it online through a platform like eBay or Etsy. Online marketplaces can reach a wider audience and have lower overhead costs than a physical store.

In conclusion, selling jewelry can be a profitable venture, but it is important to understand how to price items and where to sell them. A jewelry appraisal can help determine the value of a piece, and research and marketing can help price items competitively. There are many options to sell jewelry, including jewelers, pawnbrokers, collectors, auctions, and online marketplaces.

CHAPTER 10

Conclusion: A summary of the key points covered and resources for further information.

In this book, we have covered the basics of jewelry appraisal and its importance. We have also discussed the distinct types of appraisals, how to identify and evaluate gemstones and metals, and the various methods used to appraise jewelry. Additionally, we have explored the factors that can affect the value of jewelry and how jewelry appraisals can be used for insurance and estate planning purposes. Lastly, we have shared tips for selling jewelry.

To sum up, jewelry appraisal is a valuable tool for understanding the value and quality of jewelry pieces. It is essential for insurance, estate planning and selling purposes. The knowledge of gemstones, metals and the market comparison method, replacement value, and cost approach are important to understand the value of a piece of jewelry.

For further information, there are several resources available, including professional organizations such as the American Society of Appraisers, the International Society of Appraisers, and the National Association of Jewelry Appraisers. Additionally, books and online resources as International Gem Society (IGS) can provide more in-depth information about al topics related to gemology and jewelry appraisal topics.

CHAPTER 11

GEMOLOGIST

A gemologist is a professional in the field of gemology, which is the study of natural and artificial gemstones. imitation, synthetic and labs created. Gemologists use specialized tools and techniques to ID, evaluate, and grade gemstones based on their physical and optical properties. They may work in a variety of settings, including jewelry stores, laboratories, and research institutions, and may also be involved in lapidary, buying, selling, and **appraisal of gems, rough or faceted and jewelry**.

USPAP

USPAP are the Uniform Standards of Professional Appraisal Practice. It is a set of standards developed by The Appraisal Foundation that establishes ethical and performance standards for appraisers in the United States. USPAP is the accepted ethical and performance standards for the appraisal profession, and is required by government agencies, financial institutions, and the courts. It includes standards for appraisals of real property, individual property, and business interests. The standards cover aspects such as scope of work, qualifications of the appraiser, and the format of the appraisal report. It is important for appraisers to be familiar with and adhere to USPAP to maintain their credibility and integrity in the profession.

USPAP RULES

USPAP, the Uniform Standards of Professional Appraisal Practice, is a set of rules and guidelines that govern the appraisal profession in the United States. USPAP is to promote consistency and credibility in the appraisal profession by establishing ethical and performance standards for appraisers.

The USPAP rules cover key areas, including:

Ethics: USPAP includes a code of ethics that appraisers must abide by, which includes requirements for independence, confidentiality, and integrity.

Competency: Appraisers must have the necessary knowledge, experience, and education to complete an assignment.

Scope of work: Appraisers must clearly define the scope of their work and the intended use of the appraisal.

Appraisal report: Appraisers must prepare a written report that includes all relevant information and complies with USPAP requirements.

Record keeping: Appraisers must maintain records of their work, including all data, assumptions and methods used in their analysis.

Disclosure: Appraisers must disclose any potential conflicts of interest.

THE 5 BASIC RULES OF USPAP

1-Appraisal independence: Appraisers must be independent and unbiased in their work and must not have any conflicting interests.

2-Competency: Appraisers must have the necessary knowledge, experience, and education to complete an assignment.

3-Scope of work: Appraisers must clearly define the scope of their work and the intended use of the appraisal.

4-Appraisal report: Appraisers must prepare a written report that includes all relevant information and follows USPAP requirements.

5-Record keeping: Appraisers must keep records of their work, including all data, assumptions and methods used in their analysis.

It is worth noting that these are just the basic rules of USPAP, and the full set of standards is quite extensive and detailed. These five rules are the foundation of the USPAP standards, and they are meant to ensure that appraisals are made in a professional, unbiased, and consistent manner.

THE KIMBERLY PROCESS

The Kimberley Process Certification Scheme (KPCS) is the international initiative to prevent the circulation of conflict diamonds in international trading. This process was stablished in 2003 with the goal of ensuring that diamond purchases do not finance violence against governments and civil populations.

The process requires diamond-producing countries to supply certificates for diamonds they export, attesting that they are conflict-free. The certificates must be issued by the government of the country of origin, and must include information about the origin, weight, and other details of the diamond.

The KPCS has participation of 81 countries, and the diamonds that are certified under the KPCS are considered "conflict-free" and can be traded internationally. The KPCS also has a system of monitoring and review to ensure that participating countries are following the rules.

Critics say that the KPCS has not been effective in preventing the trade of conflict diamonds as it has some loopholes, it is not well enforced, and it does not include provisions for monitoring human rights violations in diamond mines. Despite these criticisms, the KPCS is the most effective tool to date in preventing the trade of conflict diamonds.

ABOUT THE AUTHOR

Is a Certified Gemologist (CG) and Certified Diamond Specialist (CDS) with specializations in Ruby, Sapphire, Emerald, Opal, Pearls, and Gemstones Business, is a pro member of International Gem Society (IGS).